Finding Patterns

Farm Patterns

by Nathan Olson

Capstone
press
Mankato, Minnesota

A+ Books are published by Capstone Press,
151 Good Counsel Drive, P.O. Box 669, Mankato, Minnesota 56002.
www.capstonepress.com

1 2 3 4 5 6 12 11 10 09 08 07

Library of Congress Cataloging-in-Publication Data
Olson, Nathan.
     Farm patterns / by Nathan Olson.
     p. cm.—(A+ books. Finding patterns)
     Summary: "Simple text and color photographs introduce different kinds of patterns seen on farms"—Provided by publisher.
     Includes bibliographical references and index.
     ISBN-13: 978-0-7368-6732-0 (hardcover)
     ISBN-10: 0-7368-6732-5 (hardcover)
     ISBN-13: 978-0-7368-7850-0 (softcover pbk.)
     ISBN-10: 0-7368-7850-5 (softcover pbk.)
     1. Pattern perception—Juvenile literature. 2. Farms—Miscellanea—Juvenile literature. I. Title. II. Series.
BF294.O53 2007
516'.15—dc22   2006018194

Credits

Jenny Marks, editor; Renée Doyle, designer; Charlene Deyle, photo researcher; Scott Thoms, photo editor

Photo Credits

Brand X Pictures/Bob Rashid, cover (field); Corbis/Bob Krist, 4–5; Corbis/Bryan Peterson, 15; Corbis/David Zimmerman, 26–27; Corbis/Gary W. Carter, 13, 29; Corbis/Louie Psihoyos, 8; Corbis/Steve Terrill, 14; Corbis/Tim McGuire, 7; Corbis/zefa/Malcom Fife, 9; Corbis/zefa/Robert Llewelyn, 17; Getty Images Inc./Iconica/Noah Clayton, 18; Getty Images Inc./Photographer's Choice/Richard Price, cover (barn); Getty Images Inc./Stone/Andrew Sacks, 22; Getty Images Inc./Taxi/William S. Paton, 23; Grant Heilman Photography/Arthur C. Smith III, 19; Lynn M. Stone, 25; Richard Cummins, 10; Richard Hamilton Smith, 6; Shutterstock/Dragan Trifunovic, 20; Shutterstock/Jim Parkin, 16; Shutterstock/Richard Sheppard, 21; SuperStock/age fotostock, 11, 24; SuperStock/George Hunter, 12

Note to Parents, Teachers, and Librarians

Finding Patterns uses color photographs and a nonfiction format to introduce readers to seeing patterns in the real world. *Farm Patterns* is designed to be read aloud to a pre-reader, or to be read independently by an early reader. Images and activities encourage mathematical thinking in early readers and listeners. The book encourages further learning by including the following sections: Table of Contents, Farm Pattern Facts, Glossary, Read More, Internet Sites, and Index. Early readers may need assistance using these features.

Table of Contents

What Is a Pattern?

A pattern is a repeated shape or color. Let's see what patterns we can find on the farm.

Farmers plant row after row of crops. Repeating rows make a pattern.

A field of green grass
growing on the farm
does not make a pattern.

Field and Fence Patterns

When farmers mow hay, stripes fill the field with a pattern.

Hay bales stack up in a square pattern. The twine used to tie the bales makes a pattern too.

A pattern of long fences
winds around the horse farm.

Icicles sparkle during winter months. This farm pattern is made from frozen water drops.

Some farm patterns are very big. They curve round and round as far as you can see.

Other farm patterns are as tiny as the stripes on a caterpillar.

Rows of evergreens stretch out
in a straight-line pattern on
a tree farm.

Pretty purple paths of lavender plants make a single color pattern.

Machine Patterns

Giant tractor tire treads have patterns that grip the ground.

Tracks from the tire treads press a pattern into the soft farm soil.

17

Machines water farm crops with a pattern of pipes, wheels, and hoses.

Combines cut ripe grain
from the fields. Side by side,
they work in a pattern.

Farm Animal Patterns

This pony pattern is made by horses in their stalls. You might say that they are "neigh" -bors!

Curious cows peek out of their pen in a pattern.

Hungry pink piglets eat
ear-to-ear in a pattern.

Four little barn owls line up in a pattern. Aren't they a hoot?

23

Hens lay fresh eggs in the henhouse every day. These hens nest in an orange and black pattern.

This feathered fowl rules the roost. The pointy red pattern on a rooster's head is called a comb.

Patterns of all colors and
shapes grow on the farm.

Farm Pattern Facts

Farmers don't plant their crops in rows just to make patterns. The space between the rows is for the wheels of farm machinery to move. Tractors and combines can drive through planted fields without crushing the crops growing there.

A combine is a machine farmers use to harvest crops. The machine combines several functions including cutting, threshing, and cleaning grain.

Both female chickens, called hens, and male chickens, called roosters, have combs. These red, skinlike crowns are larger on roosters than on hens. The red skin at the rooster's throat is called a waddle.

Tree farming is a way to replace forests that have been cut down. People use trees to make homes, furniture, paper, and many other products.

Sometimes crops need more rain than they can get. Farmers use irrigation machines to keep their thirsty plants watered. Irrigation machines are like giant lawn sprinklers. They can spray hundreds of gallons of water over many acres of growing crops.

Tomato hornworms are caterpillars that get their name from their red, hornlike spike. While they munch tomato stems and leaves, the pattern of stripes on their bodies helps them hide from hungry birds.

Glossary

bale (BALE)—a large bundle of straw or hay tied tightly together

barn owl (BARN OUL)—a type of owl that lives in barns and eats mice and other rodents

combine (COM-bine)—a large farm machine that is used to gather crops

fowl (FOUL)—a bird, such as a chicken, turkey, or duck

henhouse (HEN-houss)—a shelter with small spaces where hens lay their eggs

irrigation (ihr-uh-GAY-shuhn)—suppling water to crops using a system of pipes or channels

lavender (LAV-uhn-dur)—a plant with pale purple flowers that have a pleasant smell

mow (MOH)—to cut grass, grain, or hay

roost (ROOST)—a place where birds rest or build nests

tread (TRED)—a ridge on a tire that keeps the tire from slipping

Read More

Hoena, B. A. *The Farm.* Visit to... Mankato, Minn.: Capstone Press, 2004.

Macken, JoAnn Early. *Farm Animals.* Animal Worlds. Milwaukee: Gareth Stevens, 2002.

Pistoia, Sara. *Patterns.* MathBooks. Chanhassen, Minn.: Child's World, 2006.

Internet Sites

FactHound offers a safe, fun way to find Internet sites related to this book. All of the sites on FactHound have been researched by our staff.

Here's how:

1. Go to *www.facthound.com*
2. Select your grade level.
3. Type in this book ID **0736867325** for age-appropriate sites. You may also browse subjects by clicking on the letters, or by clicking on pictures and words.
4. Click on the **Fetch It** button.

FactHound will fetch the best sites for you!

Index